FUNDAMENTALS OF MANAGING CONSTRUCTION CONTRACTS

Steven Smith, Ph.D.

Wisdom Publishers

ISBN: 9798862365825
Imprint: Independently published

Cover design by: Art Painter
Library of Congress Control Number: 2018675309
Printed in the United States of America

To all the contract managers who work tirelessly to protect their organizations and ensure that contracts are fulfilled, this book is for you. I know that contract management can be a challenging and complex profession, but it is also an essential one. Contract managers play a vital role in helping organizations to achieve their goals and objectives.

This book is my way of saying thank you to all of the contract managers who work hard every day to make a difference. I hope that this book will be a valuable resource for you and that it will help you take your contract management skills to the next level.

A contract, properly drawn, is the foundation of all business.

SAMUEL GOLDWYN

CONTENTS

INTRODUCTION

E ffective construction contract management is essential to the success of any construction project. It involves the planning, execution, and administration of a construction contract in a way that minimizes risks, maximizes efficiency, and ensures that all parties involved meet their contractual obligations. This book provides a comprehensive and practical guide to effective construction contract management. It covers several aspects of the contract management process, from the pre-contract phase through to project closeout and evaluation. In addition to providing detailed guidance on each step of the contract management process, the book also includes case studies and best practices. This makes it an essential resource for any construction professional who wants to master the art of effective contract management.

Construction contracts are typically complex and high-value documents. They can involve a wide range of stakeholders, including the client, contractor, subcontractors, suppliers, and consultants. Effective contract management is essential to ensure that all parties meet their contractual duties and that the project is completed on time, within budget, and to the required quality standards. Poor contract management can lead to a number of problems, including disputes and delays, cost overruns quality defects, and legal claims and litigation. Effective contract management can help to avoid these problems and ensure that construction projects are completed successfully. Managing a construction contract effectively can provide some benefits to the parties involved in a construction project, including reduced risks, increased efficiency, improved communication and collaboration, better cost control, improved quality, enhanced stakeholder satisfaction, and increased profitability. This book is relevant to every construction professional who wants to master the art of contract management.

The book is divided into 10 chapters, which cover different aspects of the construction contract management process. Chapter 1 begins with an introduction to construction contract management, which explains the role of a construction contract manager, the importance of effective contract management, and an overview of the construction contracting industry. Chapter 2 discusses the pre-contract phase, which includes project planning and feasibility analysis, identifying stakeholders and their roles, the tendering and bidding processes, and risk assessment and mitigation strategies. Chapter 3 focuses on contract formation and documentation, which covers the different types of construction contracts, drafting effective construction contracts, contract negotiation and award, and key contract clauses and terms.

Chapter 4 addresses contract administration, which examines contract kick-off and mobilization, change order management, quality control and assurance, and progress and performance monitoring. Chapter 5 discusses communication and stakeholder management, which covers effective communication strategies, stakeholder engagement and collaboration, and conflict resolution and dispute avoidance. Chapter 6 focuses on budgeting and cost control, which explains budget development and management, cost estimation and control techniques, and tracking and managing project expenses. Chapter 7 examines legal and regulatory compliance, including understanding legal obligations, environmental and safety regulations, and compliance monitoring and reporting. Chapter 8 covers risk management, including identifying and assessing project risks, risk mitigation and contingency planning, and insurance and bonding in construction contracts. Chapter 9 discusses project closeout and evaluation, which explains final inspections and punch lists, contractual responsibilities and deliverables, and lessons learned and post-project evaluation. Chapter 10 addresses technology and tools in contract management, including construction management software, building information modeling, and automation and data analytics. Considering the significance of this book to effective contract management in the construction sector, I wish you a great read as you make the most of this resource for contract administration in your projects.

CHAPTER 1: INTRODUCTION TO CONSTRUCTION CONTRACT MANAGEMENT

Construction contract management is the process of planning, executing, and administering a construction contract in a way that minimizes risks, maximizes efficiency, and ensures that all parties involved meet their contractual duties. Construction contracts are typically complex and high-value documents. They can involve a wide range of stakeholders, including the client, contractor, subcontractors, suppliers, and consultants. Effective contract management is essential to ensure that all parties are aligned on the project goals and objectives and that the project is completed on time, within budget, and to the required quality standards. A brief overview of the key phases of construction contract management includes the following:

Pre-contract phase: This phase involves planning the project, identifying stakeholders, and developing the contract documents.

Contract formation and documentation: This phase involves negotiating and finalizing the contract documents.

Contract administration: This phase involves managing the contract throughout the construction process. This includes tasks such as monitoring progress and performance, managing change orders, and resolving disputes.

Project closeout and evaluation: This phase involves completing the project and finalizing all contractual requirements. This also includes evaluating the project and learning from any successes or failures.

Effective construction contract management requires a combination of skills and knowledge, including an understanding of construction contracts and laws, the ability to manage risks, the ability to communicate and collaborate effectively with stakeholders, strong organizational and time management skills, the ability to work independently and as part of a team. Contract managers play a vital role in the success of any construction project. By effectively managing the contract, administering the contract, managing risks, and communicating and collaborating with stakeholders, contract managers help to ensure that projects are completed successfully and meet the expectations of all parties involved.

Contract managers can be confronted with certain challenges, including challenges associated with managing complex contracts and relationships with all stakeholders, dealing with unexpected events, such as bad weather or unforeseen site conditions, balancing competing interests, such as the client's desire for a high-quality product at a low price and the contractor's desire to make a profit.

1.1 The Role of a Contract Manager

Contract managers play a vital role in the success of any

construction project. They are responsible for overseeing and managing all aspects of a construction contract, from the pre-contract phase to project closeout. Their primary goal is to ensure that the project is completed on time, within budget, and to the required quality standards. By effectively managing the contract, administering the contract, managing risks, and communicating and collaborating with stakeholders, contract managers help to ensure that projects are completed successfully and meet the expectations of all parties involved. Construction contract management is a challenging but rewarding career. Contract managers have the opportunity to make a real difference in the world by helping to build the infrastructure and facilities that we rely on every day. Contract managers typically have a wide range of responsibilities, including managing the contract, administering the contract, managing risks, and communicating and collaborating with stakeholders. Let us further discuss each of these responsibilities.

Managing the contract

Contract managers are responsible for reviewing and negotiating the contract before it is signed. This includes ensuring that the contract is fair and equitable to all parties involved and that it clearly outlines the project scope, schedule, budget, and quality standards. Once the contract is signed, the construction contract manager is responsible for managing it throughout the construction process. This includes managing change orders and resolving disputes. Change orders are requests to modify the scope of work or contract terms. Disputes can arise between any of the parties involved in the project, and the construction contract manager is responsible for resolving them in a fair and timely manner.

Administering the contract

Contract managers are responsible for administering the contract throughout the construction process. This includes managing the contract schedule and budget, monitoring progress and performance, and ensuring that all parties are meeting their contractual obligations. The construction contract manager is responsible for developing and managing the project schedule. The schedule should be realistic and achievable, and it should be updated on a regular basis to reflect changes to the project scope or other unforeseen events.

The construction contract manager is also responsible for developing and managing the project budget. The budget should be comprehensive and realistic, and it should be monitored on a regular basis to ensure that the project stays on track. The construction contract manager is responsible for monitoring the progress and performance of the project. This includes tracking the project schedule and budget, and identifying and addressing any potential problems early on. The construction contract manager is also responsible for ensuring that all parties are meeting their contractual obligations. This includes reviewing and approving contractor invoices, and taking corrective action if any party is not meeting their obligations.

Managing risks

Contract managers are responsible for identifying and assessing project risks. This includes risks such as bad weather, unforeseen site conditions, and delays in receiving materials. Once risks have been identified, the construction contract manager is responsible for developing and implementing risk mitigation strategies. These strategies may include developing contingency plans, purchasing insurance, or bonding the project.

Communicating and collaborating with stakeholders

Contract managers are responsible for communicating and collaborating effectively with all stakeholders involved in the project. This includes the client, contractor, subcontractors, suppliers, consultants, and regulatory authorities. The construction contract manager is responsible for keeping all stakeholders informed of the project's progress and performance. They are also responsible for resolving any disputes that may arise between stakeholders. The construction contract manager must be able to build and maintain positive relationships with all stakeholders. This is essential for ensuring that the project is completed successfully.

1.2 Importance of Effective Contract Management

Contract management is a critical function in any business, as contracts govern most of the relationships that a business has with its customers, suppliers, and other partners. Effective contract management can help businesses to:

Reduce risks: Contracts can be complex and legally binding documents, and if they are not managed effectively, they can expose a business to a variety of risks, such as financial losses, legal disputes, and reputational damage. Effective contract management can help to identify and mitigate these risks by ensuring that all parties to the contract understand their rights and obligations, and that the contract is properly executed and monitored.

For example: A construction company that does not effectively manage its contracts may be at risk of delays and cost overruns on its projects. This could lead to financial losses, disputes with clients, and damage to the company's reputation. By effectively managing its contracts, the construction company can reduce these risks and improve its chances of success.

Improve efficiency: Contracts can be a source of significant administrative burden for businesses. Effective contract management can help to streamline the contracting process and reduce the amount of time and resources that businesses spend on managing contracts. This can free up businesses to focus on other core activities.

For example: A company that has a large number of contracts may find it difficult to keep track of all of them and to ensure that they are all being managed effectively. By using effective contract management software, the company can streamline the contracting process and reduce the amount of time and resources that it needs to spend on managing contracts.

Increase profitability: Effective contract management can help businesses to increase their profitability by ensuring that they are getting the best possible value from their contracts. This may involve negotiating better contract terms, managing supplier performance, and identifying and recovering additional revenue opportunities.

For example: A company that is able to negotiate better contract terms with its suppliers may be able to reduce its costs. This can lead to increased profitability. Additionally, a company that effectively manages its supplier performance may be able to avoid delays and disruptions, which can also improve profitability.

Strengthen relationships with stakeholders: Contracts are the foundation of many business relationships. Effective contract management can help to strengthen these relationships by building trust and confidence between the parties involved. This can lead to improved collaboration and cooperation, which can benefit all stakeholders.

For example: A company that has a good reputation for managing its contracts is more likely to be able to attract and retain customers and suppliers. Additionally, a company that has good relationships with its stakeholders is more likely to be able to resolve disputes quickly and amicably.

Overall, effective contract management is essential for businesses of all sizes. By effectively managing their contracts, businesses can reduce risks, improve efficiency, increase profitability, and strengthen relationships with stakeholders. Effective contract management is a critical function in any business. By investing in effective contract management, businesses can reap a number of benefits, including reduced risks, increased efficiency, improved profitability, and strengthened relationships with stakeholders.

1.3 Overview of the Construction Contracting Industry

The construction contracting industry is a vast and complex industry that plays a vital role in shaping our modern world. It is a realm where creative vision, engineering precision, logistical prowess, and regulatory compliance converge to bring human dreams and necessities to life.

At the core of this industry lies a myriad of stakeholders, each playing a distinct role in orchestrating the construction process. Clients, ranging from individuals to governments and corporations, initiate projects, defining their objectives, financial parameters, and overarching vision. Contractors, the embodiment of execution, bridge the gap between ideas and reality. They assume various roles, from general contractors overseeing entire projects to specialized subcontractors responsible for specific trades. Their responsibility lies in meticulously coordinating resources, labor, and expertise to ensure that projects are not just constructed but crafted with precision. Consultants, chiefly architects and engineers, bring the creative and technical facets to life. They work in tandem with clients, taking conceptual ideas and transforming them into tangible blueprints that encompass aesthetic design, structural integrity, safety considerations, and compliance with regulatory standards.

Suppliers form the foundational backbone of the industry, providing the raw materials and equipment essential for construction. Be it lumber, concrete, steel, or advanced machinery, suppliers constitute the vital link in the supply chain that fuels the industry's engines. Regulators and governmental entities wield authority over safety, quality, and compliance. Their role is to establish, enforce, and uphold the rules, codes, and standards that construction projects must adhere to. Their oversight guarantees not only functional structures but also ensures safety and environmental responsibility. The journey of a construction project unfolds through distinct phases, each demanding meticulous attention to detail:

Planning: Clients define their needs, goals, and budget constraints. Preliminary concepts take shape, setting the stage for future development.

Design: Architects and engineers craft intricate plans and specifications that encompass everything from material selection to construction methodologies.

Bidding and contracting: Contractors present competitive bids, followed by negotiations to shape contractual terms encompassing costs, schedules, and project scope.

Construction: Contractors, subcontractors, and suppliers mobilize their resources, forging the project from conceptual vision into concrete reality.

Inspection and quality control: Regulators and inspectors ensure that construction activities meet safety and quality standards, guaranteeing the project's integrity.

Completion: The project is handed over to the client after final inspections, documentation, and last-minute adjustments ensure alignment with the client's vision and contractual obligations.

The construction contracting industry is not devoid of its own set of challenges. Cost management is an ever-present concern, demanding the delicate balance of budgets against fluctuating material and labor costs. Navigating intricate regulatory frameworks requires expertise, as each region may possess distinct requirements. Environmental considerations necessitate sustainable construction practices and materials.

Despite these challenges, the construction contracting industry is a hotbed of innovation. Technologies such as Building Information Modeling revolutionize project management and enhance efficiency. Collaborative platforms and data-driven decision-making are becoming the standard. Sustainability, once a trend, is now an imperative, with green building materials and energy-efficient designs reshaping the industry's approach.

In essence, the construction contracting industry is not merely a builder of physical spaces; it's an economic powerhouse. It generates employment, fuels economic growth, and contributes significantly to the development and enhancement of critical infrastructure. The global construction industry is valued at over $10 trillion, and it is expected to grow at a CAGR of over 5% in the coming years. The construction contracting industry is a significant contributor to the global economy, generating millions of jobs and billions of dollars in economic activity each year. It is the architect of modern progress, crafting the environments we inhabit and the infrastructure that sustains our daily lives.

The construction contracting industry is highly fragmented, with a large number of small and medium-sized businesses operating in the sector. However, there is a growing trend towards consolidation in the industry, with larger contractors acquiring smaller businesses to gain market share and expand their capabilities.

Unique Aspects of the Construction Contracting Industry

The construction contracting industry is unique in several ways. First, it is a highly project-oriented industry. Each construction project is unique, with its own set of challenges and requirements. This demands a high level of adaptability and flexibility from all stakeholders involved.

Second, the construction contracting industry is highly collaborative. Clients, contractors, consultants, suppliers, and regulators must all work together to ensure the success of a project. This requires effective communication, coordination, and problem-solving skills.

Third, the construction contracting industry is a highly regulated industry. Construction projects must adhere to a strict set of codes and standards to ensure safety, quality, and compliance. This requires a deep understanding of the relevant regulations and the ability to apply them effectively. The construction contracting industry is also facing a number of challenges, including:

A shortage of skilled labor: The construction industry is facing a growing shortage of skilled workers, due to a combination of factors such as an aging workforce, a lack of training programs, and competition from other industries.

Rising material costs: The cost of construction materials has been rising steadily in recent years, due to factors such as supply chain disruptions and increased demand from emerging markets.

Increased regulatory compliance: The construction industry is subject to a complex and evolving set of regulations, which can add to the cost and complexity of projects.

Despite these challenges, the construction contracting industry is a vital and growing sector of the global economy. It is an industry that is essential to our modern way of life, and it is an industry that is constantly innovating to meet the needs of a changing world.

Emerging Trends in the Construction Contracting Industry

The construction contracting industry is constantly evolving, and there are a number of emerging trends that are shaping the future of the sector. One of the most significant trends is the rise of digital technologies. Technologies such as Building Information Modeling, virtual reality, and augmented reality are being used to improve project planning, coordination, and execution.

Another important trend is the growing focus on sustainability. Construction firms are increasingly looking for ways to reduce the environmental impact of their projects, through the use of green building materials, energy-efficient designs, and waste reduction strategies. The construction contracting industry is also becoming more globalized, with contractors increasingly competing for projects in international markets. This is leading to a growing demand for cross-cultural collaboration and expertise.

CHAPTER 2: PRE-CONTRACT PHASE

Contract managers leverage the pre-contract phase to lay the foundation for successful project execution. This involves fostering collaboration, establishing clear communication, managing expectations, and mitigating risks.

2.1 Project Planning and Feasibility Analysis

Project planning and feasibility analysis are essential components of any successful construction project. Contract managers play a critical role in both of these areas, as they are responsible for developing and executing the project plan, and for ensuring that the project is feasible and viable.

Project Planning

Project planning is the process of developing a roadmap for the project, outlining the steps that need to be taken to achieve the desired outcome. A well-developed project plan will help to ensure that the project is completed on time, within budget, and to the required quality standards. Contract managers should begin the project planning process by developing a clear understanding of the client's needs and objectives. Once this has been done, they can begin to develop a detailed plan that outlines the following:

Scope of work: The scope of work should define all of the work that needs to be completed in order to deliver the project to the client's satisfaction. It is important to be as specific as possible when defining the scope of work, as this will help to avoid any misunderstandings or disputes down the road.

Schedule: The project schedule should outline the start and end dates for each task, as well as the dependencies between tasks. It is important to be realistic when developing the project schedule, as any delays in one task can have a cascading effect on the rest of the project.

Budget: The project budget should estimate the cost of each task, as well as the overall cost of the project. It is important to be conservative when developing the project budget, as there are often unforeseen costs that can arise during the course of a construction project.

Resources: The project plan should identify the resources that will be needed to complete the project, such as labor, equipment, and materials. It is important to ensure that the necessary resources are available before the project begins, as any shortages can lead to delays and cost overruns.

Risk management: The project plan should identify and assess potential risks, and develop mitigation strategies for each risk. It is important to regularly monitor and update the risk management plan, as new risks may emerge during the course of the project.

Feasibility Analysis

Feasibility analysis is the process of evaluating the viability of a construction project. It involves assessing the technical, economic, and environmental feasibility of the project. Contract managers should conduct a feasibility analysis before any construction activities begin. This will help to ensure that the project is feasible and that there are no insurmountable challenges that could prevent it from being completed successfully. The following factors should be considered when conducting a feasibility analysis:

Technical feasibility: This involves assessing the technical feasibility of the project, such as the availability of the necessary materials and equipment, and the expertise of the project team.

Economic feasibility: This involves assessing the economic feasibility of the project, such as the cost of the project and the potential return on investment.

Environmental feasibility: This involves assessing the environmental impact of the project, and ensuring that it complies with all applicable environmental regulations. Contract managers can consider these additional measures to produce a robust planning and feasibility analysis.

Involve all stakeholders early in the process: It is important to involve all stakeholders in the project planning and feasibility analysis process as early as possible. This will help to ensure that everyone is on the same page and that all perspectives are considered.

Use data-driven decision-making: When making decisions about project planning and feasibility, it is important to rely on data as much as possible. This data can be used to develop realistic schedules and budgets, and to assess the risks and potential rewards of the project.

Be flexible and adaptable: Things don't always go according to plan in the construction industry. It is important to be flexible and adaptable when managing projects. Be prepared to make changes to the project plan as needed, and be prepared to deal with unexpected challenges.

2.2 Identifying Stakeholders and Their Roles

Construction contract management starts with a rather complex task - figuring out who's involved and what they do in a project. It's a bit like unwrapping the layers of a complicated story to discover the main characters and their special roles. At the heart of any construction project is the client, the one who starts it all. Clients can be individuals wanting to build their dream homes or big organizations, like the government, working on major projects. Their role is crucial; they explain what they want to achieve, how much they can spend, and what the final project should look like.

Now, contractors are the ones who make things happen in this big production. General contractors oversee everything, like a conductor directing an orchestra, ensuring all the different parts come together. Subcontractors, the specialists, handle specific tasks like plumbing or electrical work, making sure everything matches the client's vision. Then, we have the consultants, architects, and engineers. Think of them as the designers and structural experts. Architects make things look great, while engineers make sure everything is strong, safe, and follows the rules. Suppliers are like the providers, bringing in all the necessary materials and tools. They make sure everything arrives on time and is of good quality, sort of like the props and costumes for our big production. Regulators play a critical role, like referees in a game. They make sure everyone follows the rules, ensuring the project is safe and follows the law. Now, let's understand what these roles mean:

Clients share their vision, budget, and expectations with the Contract Manager.

Contractors are the doers, turning the client's ideas into real structures.

Consultants, like architects and engineers, design and ensure everything is safe and within the rules.

Suppliers bring in the materials and tools needed for the construction.

Regulators ensure everyone follows the rules to keep the project safe and legal.

Consider the Contract Manager as the director of a big play. They make sure all these different characters work together smoothly to create a successful project, a bit like making sure everyone on stage plays their part right. They also plan for potential problems and make sure everything is done just right to make the project a success.

2.3 Tendering and Bidding Processes

In various industries, including construction, "tendering" and "bidding" are fundamental processes often used to find the right people or companies for a project. To simplify these concepts, let's break them down into easily understandable terms.

Tendering: Requesting Proposals

Think of tendering as an invitation to get proposals or bids. Imagine you want to build a new house, and you're looking for someone to do it. You don't just pick the first builder you find. Instead, you want to ensure you find the best fit for your project. So, you create a clear list of what you want in your house. This list is known as "tender documents."

These documents include everything from your desired design to the materials you'd like to use and the timeline for completing the project. It's like providing builders with a detailed recipe to follow.

Bidding: Providing Proposals

On the other side of the equation are the builders or contractors. They receive your "invitation" in the form of tender documents and then start preparing their "proposals" or "bids." These proposals outline their detailed plans for how they will construct your house, including the cost and estimated project completion date.

Builders don't just consider the cost; they also think about how they can deliver quality within your budget. They might suggest certain materials or construction methods. So, bidding isn't solely about offering the lowest price; it's about demonstrating their ability to perform the job excellently within your budget.

Contract Managers: The Facilitators

Consider yourself as the person responsible for making all of this happen. You act as the intermediary ensuring that everything proceeds smoothly. These individuals are Contract Managers. They assist in creating clear and equitable tender documents for clients, making sure they encompass all crucial details.

Contract Managers also help clients in selecting the best proposal, considering factors such as the builder's experience, reputation, and how closely their proposal aligns with the client's requirements. They also aid in managing potential risks, such as unforeseen issues during construction, and ensure that all involved parties comprehend their roles and responsibilities.

Bringing it All Together

In essence, tendering and bidding are akin to matchmaking processes. The client (the entity seeking to have something constructed) sends out an invitation with explicit instructions (tender documents). The builders (contractors) craft their plans and cost estimates (proposals or bids). The Contract Manager acts as the matchmaker, aiding the client in choosing the most suitable builder for the project and guaranteeing that everything proceeds smoothly.

When you encounter discussions about tendering and bidding, consider it as a method of finding the right individual or company to bring your project to fruition, whether it's constructing a house, a bridge, or any other undertaking. It's a structured approach to connect projects with capable builders, like a well-organized matchmaking process for construction projects!

2.4 Risk Assessment and Mitigation Strategies

Contract managers play a critical role in identifying, assessing, and mitigating risks in the context of contracts. By understanding the different types of risks that can arise in a contract, and by developing and implementing effective mitigation strategies, contract managers can help to ensure that contracts are executed successfully and that the interests of all parties are protected.

Types of Risks in Contracts

There are a variety of different types of risks that can arise in a contract, including:

Performance risk: This risk refers to the possibility that one or more parties to the contract will fail to perform their obligations under the contract. Performance risk can be caused by a variety of factors, such as financial difficulties, technical problems, or managerial incompetence.

Schedule risk: This risk refers to the possibility that the contract will not be completed on time. Schedule risk can be caused by a variety of factors, such as unforeseen delays, bad weather, or resource constraints.

Budget risk: This risk refers to the possibility that the contract will exceed its budget. Budget risk can be caused by a variety of factors, such as unforeseen costs, changes to the scope of work, or poor cost estimation.

Quality risk: This risk refers to the possibility that the products or services delivered under the contract will not meet the required quality standards. Quality risk can be caused by a variety of factors, such as poor workmanship, defective materials, or inadequate quality control procedures.

Compliance risk: This risk refers to the possibility that the contract will not be in compliance with applicable laws and regulations. Compliance risk can be caused by a variety of factors, such as inadequate legal review of the contract, changes in the law, or failure to obtain required permits.

Risk Assessment

The first step in mitigating risks in a contract is to identify and assess the risks that are present. This can be done by conducting a risk assessment. A risk assessment involves identifying all of the potential risks that could arise in the contract, and then assessing the likelihood of each risk occurring and the potential impact of each risk if it does occur. Once the risks have been identified and assessed, the contract manager can develop and implement mitigation strategies to reduce the likelihood of the risks occurring and to minimize the impact of the risks if they do occur.

Risk Mitigation Strategies

There are a variety of different risk mitigation strategies that can be used in contracts. Some of the most common risk mitigation strategies include:

Contractual provisions: Many risks can be mitigated through contractual provisions. For example, performance risk can be mitigated by including provisions in the contract that specify the performance standards that must be met and the consequences for failing to meet those standards. Schedule risk can be mitigated by including provisions in the contract that specify the deadline for completing the contract and the consequences for failing to meet that deadline. Budget risk can be mitigated by including provisions in the contract that specify the budget for the contract and the process for managing changes to the budget. Quality risk can be mitigated by including provisions in the contract that specify the quality standards that must be met and the inspection and testing procedures that will be used. Compliance risk can be mitigated by including provisions in the contract that require the parties to comply with all applicable laws and regulations.

Insurance: Insurance can be used to mitigate a variety of risks in contracts. For example, performance risk can be mitigated by purchasing performance bonds or surety bonds. Schedule risk can be mitigated by purchasing delay-in-completion insurance. Budget risk can be mitigated by purchasing cost-overrun insurance. Quality risk can be mitigated by purchasing product liability insurance. Compliance risk can be mitigated by purchasing environmental liability insurance or other types of liability insurance.

Risk transfer: Risk transfer involves transferring the risk to another party. For example, performance risk can be transferred to a subcontractor. Schedule risk can be transferred to a supplier. Budget risk can be transferred to a customer. Quality risk can be transferred to a third-party inspector or testing laboratory. Compliance risk can be transferred to a consultant or other professional who can provide advice on compliance issues.

Risk retention: Risk retention involves keeping the risk within the organization. This is often the most cost-effective approach, but it also involves the greatest amount of risk.

The Role of Contract Managers in Risk Mitigation

Contract managers play a critical role in risk mitigation in a variety of ways. First, contract managers are responsible for identifying and assessing risks in contracts. Second, contract managers are responsible for developing and implementing risk mitigation strategies. Third, contract managers are responsible for monitoring and managing risks throughout the life of the contract.

Effective contract managers are proactive in their approach to risk mitigation. They identify and assess risks early in the contracting process, and they develop and implement mitigation strategies before those risks have a chance to materialize. They also monitor and manage risks throughout the life of the contract, and they make adjustments to the mitigation strategies as needed. The following are some specific measures that contract managers can do to mitigate risks:

- Conduct thorough due diligence on potential contractors. This includes reviewing the contractor's financial statements, references, and past performance.
- Negotiate clear and concise contracts that clearly define the scope of work, the performance standards, and the consequences for non-performance.
- Establish regular communication with the contractor and monitor the progress of the contract closely.
- Have a plan in place for dealing with unexpected events and contingencies.

- Purchase insurance to protect against unforeseen losses.

Contract managers can also mitigate risks by considering the following:

- Be aware of the latest trends and developments in risk management. This includes staying up-to-date on new laws and regulations, as well as new risk management techniques and technologies.

- Get professional help when needed. If you are not comfortable assessing or mitigating certain types of risks, don't hesitate to consult with a risk management expert.

- Keep your stakeholders informed. It is important to keep your stakeholders informed of the risks involved in a contract and the steps you are taking to mitigate those risks. This will help to build trust and confidence, and it will make it easier to deal with problems if they arise.

CHAPTER 3:
CONTRACT FORMATION AND DOCUMENTATION

C ontract formation and documentation are essential aspects of the contract management process. Contract formation is the process of creating a legally binding agreement between two or more parties. Contract documentation is the process of recording the terms of the contract in a written document. Contract managers play a critical role in both contract formation and documentation. They are responsible for ensuring that contracts are formed in a legally valid manner and that they are accurately and comprehensively documented.

3.1 Types of Construction Contracts

Contract managers play a critical role in the selection and management of construction contracts. By understanding the different types of construction contracts available and their respective advantages and disadvantages, contract managers can help to ensure that the right contract is selected for each project and that the contract is managed effectively to ensure a successful outcome. There are four main types of construction contracts:

- **Lump-sum contracts**: Lump-sum contracts are the most common type of construction contract. Under a lump-sum contract, the contractor agrees to complete the project for a fixed price. Lump-sum contracts are simple and straightforward, but they can be risky for contractors if unforeseen costs arise during the course of the project.

- **Unit price contracts**: Unit price contracts are used when the scope of work is not fully defined or when the quantities of work involved are difficult to estimate. Under a unit price contract, the contractor is paid a fixed price for each unit of work completed. Unit price contracts can be less risky for contractors than lump-sum contracts, but they can also be more complex and time-consuming to administer.

- **Cost-plus contracts**: Cost-plus contracts are used when the scope of work is not fully defined or when the quantities of work involved are difficult to estimate. Under a cost-plus contract, the contractor is reimbursed for all of their costs, plus a fee for their profit and overhead. Cost-plus contracts are the least risky for contractors, but they can also be the most expensive for project owners.

- **Time and materials contracts**: Time and materials contracts are used when the scope of work is not fully defined or when the quantities of work involved are difficult to estimate. Under a time and materials contract, the contractor is paid for their time and materials at an agreed-upon rate. Time and materials contracts can be the most flexible for project owners, but they can also be the most expensive if the project takes longer than expected to complete.

Selecting the Right Construction Contract

The type of construction contract that is selected for a particular project will depend on a variety of factors, including the scope of work, the budget, the risk tolerance of the project owner, and the experience and expertise of the contractor. Contract managers should carefully consider all of these factors before selecting a construction contract. They should also consult with the project owner and the contractor to get their input on the best type of contract for the project.

Managing Construction Contracts Effectively

Once a construction contract has been selected, the contract manager is responsible for managing the contract effectively. This includes overseeing the contractor's performance, ensuring that the project is completed on time and within budget, and resolving any disputes that may arise. Contract managers can use a variety of tools and techniques to manage construction contracts effectively. These tools and techniques include:

- **Project management software**: Project management software can be used to track the progress of the project, identify potential problems, and communicate with all stakeholders.

- **Risk management tools**: Risk management tools can be used to identify, assess, and mitigate risks to the project.

Dispute resolution tools: Dispute resolution tools can be used to resolve disputes between the project owner and the contractor.

Contract managers should also maintain good communication

with all stakeholders throughout the course of the project. This includes communicating regularly with the project owner, the contractor, and other stakeholders such as subcontractors, suppliers, and inspectors.

3.2 Drafting Effective Construction Contracts

Contract managers play a critical role in the drafting of effective construction contracts. By understanding the key elements of a construction contract and by using clear and concise language, contract managers can help to ensure that contracts are fair and enforceable, and that they accurately reflect the intentions of the parties.

Key Elements of a Construction Contract

A well-drafted construction contract should include the following key elements:

- **Scope of work**: The scope of work should clearly and concisely describe the work that the contractor is required to perform. The scope of work should also include any specific requirements or specifications that the contractor must meet.
- **Timeline**: The contract should specify a start date and a completion date for the project. The contract should also include any milestones that the contractor must meet along the way.

- **Budget**: The contract should specify the total budget for the project. The contract should also include a breakdown of the budget by category, such as labor, materials, and equipment.

- **Payment terms**: The contract should specify how and when the contractor will be paid. The contract should also include any penalties for late payment.

- **Dispute resolution**: The contract should specify how disputes between the project owner and the contractor will be resolved. The contract should also include any mediation or arbitration clauses.

- **Termination**: The contract should specify how the contract can be terminated by either party. The contract should also include any notice requirements for termination.

Drafting Clear and Concise Contracts

Contract managers play a critical role in the drafting of effective construction contracts. By understanding the key elements of a construction contract and by using clear and concise language, contract managers can help to ensure that contracts are fair and enforceable and that they accurately reflect the intentions of the parties. When drafting a construction contract, it is important to use clear and concise language. Avoid using legal jargon and technical terms that the parties may not understand. Instead, use plain English that is easy to read and understand. It is also important to make sure that the contract is well-organized and easy to follow. Use headings and subheadings to break up the text and make it easier to find specific information.

Before finalizing the contract, it is important to review it with all stakeholders, including the project owner, the contractor, and any other relevant parties. This will help to identify and address any potential problems early on. There are a number of standard contract forms available for construction projects. These forms can be a good starting point for drafting a contract. However, it is important to review the standard form carefully and make any necessary changes to reflect the specific needs of the project. Before signing the contract, it is important to have it reviewed by an attorney. An attorney can help to ensure that the contract is fair and enforceable and that it protects the interests of the project owner.

The more specific the contract is, the less likely it is that there will be disputes later on. When describing the scope of work, be as specific as possible about the materials to be used, the construction methods to be employed, and the quality standards to be met. Things don't always go according to plan in construction projects. It's important to be flexible and allow for some changes to the contract as needed. However, it's also important to protect the project owner's interests by including provisions in the contract that limit the scope of changes and the impact of changes on the budget and timeline. The contract should be fair to both the project owner and the contractor. It's important to strike a balance between protecting the project owner's interests and giving the contractor the flexibility, they need to complete the project successfully.

3.3 Contract Negotiation and Award

Contract managers play a critical role in the contract negotiation and award process. They are responsible for representing the interests of the organization and ensuring that the best possible contract is awarded to the right contractor. To achieve this, contract managers must have a deep understanding of the different types of negotiation styles, effective negotiation strategies, and the principles of fair and transparent contract award.

Types of Negotiation Styles

There are two main types of negotiation styles: competitive negotiation and collaborative negotiation.

- **Competitive negotiation**: Competitive negotiation is a win-lose negotiation style. The goal of competitive negotiation is to get the best possible deal for yourself, even if it means that the other party gets a worse deal. Competitive negotiation is often used in situations where the parties have conflicting interests and there is a limited amount of resources available.

- **Collaborative negotiation**: Collaborative negotiation is a win-win negotiation style. The goal of collaborative negotiation is to find a mutually agreeable solution that meets the needs of both parties. Collaborative negotiation is often used in situations where the parties have shared interests and there is a potential for long-term collaboration.

Effective Negotiation Strategies

Effective contract managers use a variety of negotiation strategies to achieve their desired outcomes. Some common negotiation strategies include:

- **Preparation**: The key to successful negotiation is preparation. Contract managers should carefully prepare for negotiations by understanding their own needs and interests, the needs and interests of the other party, and the strengths and weaknesses of their own position. This includes gathering information about the contractor's qualifications, experience, and financial stability.

- **Communication**: Communication is essential for effective negotiation. Contract managers should be clear and concise in their communication, and they should be willing to listen to the other party's concerns. They should also be prepared to explain their own needs and interests in a way that is persuasive and respectful.

- **Problem-solving**: Contract managers should focus on problem-solving during negotiations. They should work with the other party to find solutions that meet the needs of both parties. This may involve brainstorming creative solutions, making concessions, or finding common ground.

- **Relationship building**: Contract managers should build relationships with the other party during negotiations. This will help to create a climate of trust and cooperation, which can lead to more successful negotiations. Relationship building can be done by getting to know the other party on a personal level, finding common interests, and showing respect for their needs and concerns.

Fair and Transparent Award Process

Once negotiations have been completed, contract managers must follow a fair and transparent award process. This includes:

- **Evaluating proposals fairly and impartially**: Contract managers should evaluate proposals fairly and impartially, based on the criteria that were established in the tender documents. This criterion should be objective and measurable, and it should be communicated to all potential contractors in advance.

- **Selecting the winning bidder**: Contract managers should select the winning bidder based on the criteria that were established in the tender documents. This may involve ranking the bidders based on their scores on each criterion, or it may involve a more complex process such as a multi-criteria decision analysis.

- Awarding the contract: Contract managers should award

the contract to the winning bidder in a timely and efficient manner. This includes notifying the winning bidder of their selection and providing them with the contract documents to sign.

Contract managers play a critical role in the contract negotiation and award process. By understanding the different types of negotiation styles, developing effective negotiation strategies, and following a fair and transparent award process, contract managers can help to ensure that the right contract is awarded to the right contractor at the best possible price. One of the best ways to get a good deal in a negotiation is to be prepared to walk away. If the other party is not willing to meet your needs, you should be willing to walk away from the negotiation. This does not mean that you have to be rude or aggressive, but it does mean that you should be prepared to end the negotiation if you are not getting what you need.

If you are not comfortable negotiating on your own, don't be afraid to ask for help from a more experienced negotiator. This could be someone from your own organization, such as a senior manager or a legal advisor, or it could be an external consultant. It is important to document everything in the negotiation process. This includes keeping track of all communications, proposals, and agreements. This documentation will be helpful if any disputes arise later on.

3.4 Key Contract Clauses and Terms

Contract managers play a critical role in understanding and managing key contract clauses and terms. They are responsible for ensuring that contracts are drafted, negotiated, and managed in a way that protects the interests of their organization. Key contract clauses and terms are the essential provisions that govern the relationship between the parties to a contract. They should be drafted clearly and concisely, and they should be tailored to the specific needs of the contract. Some of the most common key contract clauses and terms include:

- **Scope of work**: This clause describes the work that the contractor is required to perform under the contract. It should be as specific and detailed as possible, to avoid any misunderstandings or disputes down the road. The scope of work clause should also include any deliverables that the contractor is required to provide.

- **Timeline**: This clause specifies the start date and end date for the project. It may also include milestones or deadlines for specific tasks. It is important to be realistic when setting the timeline, and to allow for some flexibility in case of unforeseen delays.

- **Budget**: This clause specifies the total cost of the project. It may also include a breakdown of costs by category, such as labor, materials, and equipment. It is important to create a realistic budget and to monitor costs throughout the project to ensure that the budget is not exceeded.

- **Payment terms**: This clause specifies how and when the contractor will be paid. It may include a schedule of payments, as well as any penalties for late payment. It is important to establish clear payment terms to avoid any disputes between the parties.

- **Termination**: This clause specifies how the contract can be terminated by either party. It should also include any notice requirements for termination. It is important to have a termination clause in place in case things do not go according to plan.

- **Warranties**: This clause sets out the contractor's warranties with respect to the work and the materials used. It should specify the duration of the warranties and the remedies available to the project owner if the contractor fails to meet its warranty obligations. Warranties can help to protect the project owner from financial losses if the work does not meet the agreed-upon standards.

- **Limitation of liability**: This clause limits the contractor's liability for damages arising out of the contract. It is important to note that some courts may not uphold limitation of liability clauses, so it is important to have the clause reviewed by an attorney before signing the contract.

- **Dispute resolution**: This clause specifies how disputes between the parties to the contract will be resolved. It may include a provision for mediation or arbitration, or it may specify that disputes will be resolved in court. It is important to have a dispute resolution clause in place to ensure that any disputes are resolved quickly and efficiently.

In addition to these key clauses and terms, there are a number of other clauses and terms that may be included in a contract, depending on the specific needs of the parties. For example, a contract may include clauses on confidentiality, intellectual property rights, and force majeure.

Contract managers play a critical role in understanding and managing key contract clauses and terms. They are responsible for ensuring that contracts are drafted in a way that protects the interests of their organization, and they are also responsible for negotiating and managing contracts effectively.

For contract managers to manage key contract clauses and terms, the first step is to identify the key contract clauses and terms that are relevant to the specific contract. This can be done by reviewing the contract carefully and by considering the specific needs of the organization. Once the key contract clauses and terms have been identified, it is important to understand their impact on the organization. This includes understanding the risks and liabilities associated with each clause or term. Contract managers are responsible for negotiating the key contract clauses and terms on behalf of their organization. This involves working with the other party to reach an agreement that is fair and that protects the interests of both parties. Once the contract has been signed, contract managers are responsible for managing the key contract clauses and terms. This includes monitoring the contractor's performance, ensuring that the contractor is meeting its obligations under the contract, and resolving any disputes that may arise.

CHAPTER 4:
CONTRACT
ADMINISTRATION

C ontract administration is a critical function for any organization that relies on contracts to conduct its business. By effectively managing contracts, organizations can reduce risk, ensure that they are getting the best value for their money, and build strong relationships with their suppliers. Effective contract administration requires a deep understanding of the different types of contracts, the risks involved, and the best practices for managing them. It also requires strong communication and negotiation skills. Contract administrators play a vital role in helping organizations to achieve their business goals. By effectively managing contracts, they can help to reduce costs, improve efficiency, and mitigate risk.

4.1 Contract Kick-Off and Mobilization

Contract kick-off and mobilization are the initial phases in the lifecycle of any construction project. These stages are analogous to the preparation and rehearsal period for a grand theatrical performance. Contract Managers are the directors of this performance, responsible for setting the stage and ensuring that all the actors know their roles, delivering a stellar opening act.

Contract Kick-Off: Laying the Groundwork

Contract kick-off is akin to the script reading session where all the actors gather to understand their roles and the plot. In this phase, Contract Managers play a pivotal role in ensuring that all parties are aligned and well-prepared for what lies ahead. First and foremost, a comprehensive contract review is conducted. This involves a deep dive into the contract's intricacies, deciphering legal terminology, understanding terms and conditions, and ensuring that every clause is thoroughly comprehended by all involved parties. Stakeholder alignment is another critical aspect. Contract Managers serve as communication bridges, facilitating open dialogues and resolving any initial concerns or uncertainties among stakeholders. It's essential that everyone shares a common understanding of project expectations.

Project planning is a key responsibility during contract kick-off. Contract Managers collaborate with project managers and other stakeholders to develop a robust project plan. This plan outlines timelines, milestones, resource allocation, and strategies for risk mitigation. It serves as the guiding roadmap for the project's execution. Risk assessment is an integral part of contract kick-off. Contract Managers, in collaboration with stakeholders, identify potential risks and challenges specific to the project. These risks can range from environmental factors to unforeseen delays in material deliveries or regulatory changes. The aim is to be proactive in identifying potential hurdles.

Procurement planning, if applicable, is initiated during this phase. Contract Managers oversee the selection of suppliers, negotiate contracts, and ensure that the procurement process aligns with project objectives and budgets. Transparent and effective communication is paramount throughout contract kick-off. Contract Managers are responsible for disseminating vital information, ensuring that all stakeholders are well-informed, and facilitating ongoing dialogue to maintain transparency and alignment.

Mobilization: Turning Plans into Action

As contract kick-off concludes, the project proceeds into the mobilization phase, akin to the commencement of rehearsals where actions translate script into performance. Resource allocation is a critical focus during this phase. Contract Managers collaborate closely with project managers to ensure that necessary resources, whether labor, materials, or equipment, are allocated as per plan. They oversee the procurement process to ensure that all resources are procured on time and within budget.

Compliance oversight becomes crucial during mobilization. Contract Managers diligently monitor compliance with contractual agreements and regulatory standards at every stage of the project. This is essential to safeguard the integrity of the project. Risk management continues to be a top priority. Risks identified during contract kick-off are continually monitored. Contract Managers remain vigilant, ready to activate mitigation strategies should any risks begin to materialize.

Issue resolution is another significant responsibility during mobilization. Challenges and unforeseen issues are an inherent part of any project. Contract Managers act as problem solvers, collaborating with stakeholders to resolve issues promptly, minimizing disruptions to project progression. Performance evaluation is an ongoing process during mobilization. Contract Managers closely monitor project performance, assessing whether milestones are being met, resources are being used efficiently, and quality standards are being upheld. Thorough documentation is integral. Contract Managers maintain comprehensive records of all project-related activities, ensuring transparency and accountability at every step of the project.

4.2 Change Order Management

Change order management is a multifaceted and pivotal component within the domain of contract management, particularly in the intricate landscape of construction projects. It serves as the mechanism for navigating the ever-evolving terrain of project alterations and modifications. In the construction industry, change orders represent contractual amendments that alter various aspects of a project, encompassing scope, schedule, and budget. These alterations can arise from a multitude of factors, including design adjustments, unforeseen site conditions, client-initiated modifications, or regulatory changes. As the dynamics of a construction project are seldom static, the management of change orders becomes a fundamental practice.

At the heart of efficient change order management lies the role of Contract Managers, who are akin to seasoned navigators steering a ship through shifting seas. Their responsibilities encompass an array of critical tasks, beginning with the early detection of potential changes. Effective contract managers remain vigilant throughout the project's lifecycle, collaborating closely with project managers, engineers, and other stakeholders to identify any issues or conditions that may necessitate change orders. This proactive stance allows them to anticipate and prepare for potential alterations. Accurate and meticulous documentation is paramount in the change order management process. Contract managers are meticulous record-keepers, ensuring that all change requests are comprehensively documented. These records provide a clear and detailed account of the proposed changes, the reasoning behind them, and their potential ramifications on the project's scope, schedule, and budget. The precision and comprehensiveness of these documents are vital in facilitating subsequent steps in the change order management process.

Each change request undergoes a rigorous assessment, and Contract managers are at the forefront of this evaluation. They meticulously examine the feasibility of the proposed changes, the associated costs, and the potential impacts on project timelines. This assessment is a collaborative effort, involving close coordination with project teams to determine the most efficient and cost-effective methods for implementing the requested changes. Negotiation skills are paramount in the realm of change order management. Contract managers engage in dialogues and negotiations with clients, contractors, and other stakeholders to reach mutually agreeable terms for the change orders. These negotiations often involve discussions on costs, schedules, and any necessary contractual adjustments. The ability to navigate these negotiations effectively is essential in securing the best outcomes for all parties involved.

Once agreements are reached, Contract managers ensure that all changes are diligently documented in writing. This documentation includes the revision of contract terms, project timelines, and budgets to reflect the approved change orders accurately. Clarity and transparency in this documentation are crucial to prevent potential disputes and to ensure that all parties are aligned with the agreed-upon changes. Effective communication is the linchpin of successful change order management. Contract managers shoulder the responsibility of disseminating pertinent information to the entire project team, ensuring that every member is well-informed regarding the approved changes and their implications. This transparent communication fosters an environment of accountability and ensures that everyone is aware of their roles and responsibilities in implementing the changes.

Change orders are seamlessly integrated into the broader project management processes under the watchful eye of Contract managers. Collaborating with project managers, they ensure that project plans, schedules, and budgets are updated to reflect the approved changes accurately. This integration is crucial in maintaining the project's trajectory and ensuring that it remains aligned with its goals, even in the face of modifications.

The proactive approach of Contract managers extends to risk assessment. They continuously evaluate potential risks associated with change orders. This assessment includes an examination of how the changes may impact project timelines, budgets, and the quality of the final deliverables. Identifying and addressing new risks that may emerge due to the changes is part of their ongoing responsibilities. In cases where disputes arise over change orders, Contract managers step into the role of mediators and problem solvers. Their negotiation and conflict resolution skills come into play as they work towards equitable solutions. Resolving disputes promptly and effectively is vital to prevent disruptions to the project's progression and maintain positive working relationships among stakeholders.

Even after change orders are approved and integrated into the project, Contract managers continue to monitor their impact. They assess whether the changes are being implemented as agreed upon, tracking any deviations from the approved terms, and taking corrective actions if necessary. This vigilant oversight ensures that the project remains on course and that the approved changes are executed in a manner that aligns with project objectives. Change orders, when managed efficiently, enable projects to adapt to evolving circumstances while upholding transparency, accountability, and contractual obligations. Contract managers serve as the compass, ensuring that these changes are navigated with precision. Their expertise is instrumental in mitigating risks, averting disputes, and steering projects toward successful completion, regardless of the challenges that may arise during the journey.

4.3 Quality Control and Assurance

Quality control and assurance (QA/QC) are essential components of effective contract management. QA/QC processes help to ensure that the work performed by contractors meets the agreed-upon standards and requirements. This can help to reduce risk, improve efficiency, and increase customer satisfaction. Contract managers play a critical role in overseeing QA/QC activities. They are responsible for developing and implementing QA/QC plans, monitoring the contractor's performance, and identifying and resolving any quality issues.

Contract managers work with stakeholders to develop QA/QC plans that are tailored to the specific contract. These plans should identify the quality standards and requirements that must be met, as well as the processes and procedures that will be used to assess quality. The QA/QC plan should be developed early in the contract lifecycle, and it should be reviewed and updated regularly to ensure that it remains aligned with the project's needs. The QA/QC plan should be documented and communicated to all stakeholders, including the contract manager, project manager, contractor, and any other individuals who will be involved in the project.

Contract managers monitor the contractor's performance to ensure that they are meeting the agreed-upon quality standards. This may involve conducting inspections, reviewing test results, and auditing the contractor's quality management system.

Inspections are a common way to assess the quality of the contractor's work. Inspections can be conducted at any point during the project lifecycle, but they are typically conducted at key milestones, such as the completion of a major phase of work or the delivery of a significant product or service.

Test results can also be used to assess the quality of

the contractor's work. Tests can be conducted on materials, components, or systems to ensure that they meet the agreed-upon standards. Audits of the contractor's quality management system can also be used to assess the quality of the contractor's work. Audits involve a systematic review of the contractor's quality management processes and procedures to ensure that they are effective and that they are being followed.

If quality issues are identified, contract managers work with the contractor to identify the root cause of the problem and develop a corrective action plan. This may involve reworking the work, providing additional training to the contractor's staff, or changing the project plan. It is important to identify and resolve quality issues as early as possible. This will help to minimize the impact of the quality issues on the project schedule and budget. It is also important to document all quality issues and the corrective actions that were taken. This documentation will be helpful in resolving any disputes that may arise and in improving the QA/QC process for future projects. These are some specific examples of QA/QC activities that can be implemented in different types of contracts:

- **Construction contracts**: Contract managers can implement QA/QC in construction contracts by requiring the contractor to submit shop drawings and material samples for approval, conducting regular inspections of the work, and reviewing quality control reports from the contractor's quality management system.

- **Software development contracts**: Contract managers can implement QA/QC in software development contracts by requiring the contractor to follow a defined development methodology, conduct unit testing, integration testing, and system testing, and submit the software for testing by an independent third party.

- **Information technology outsourcing contracts**: Contract managers can implement QA/QC in IT outsourcing contracts by requiring the contractor to meet specific service level agreements (SLAs), conduct regular audits of the contractor's service delivery, and review customer satisfaction surveys.

By effectively implementing QA/QC processes, contract managers can help to ensure that contracts are executed successfully and that the organization receives the highest quality of goods or services. Effective QA/QC processes require the involvement of all stakeholders, including the contract manager, project manager, contractor, and any other individuals who will be involved in the project. It is also important to tailor the QA/QC plan to the specific contract and to use a variety of QA/QC methods.

4.4 Progress and Performance Monitoring

Progress and performance monitoring are essential components of effective contract management. By tracking the progress of the contractor's work and assessing their performance, contract managers can identify and address any potential problems early on, and ensure that the contract is completed on time, within budget, and to the required quality standards.

The Importance of Progress and Performance Monitoring

There are a number of reasons why progress and performance monitoring are so important in contract management. First, it allows contract managers to identify and address any potential problems early on. This is important because it can help to prevent these problems from causing significant delays, cost overruns, or quality issues. Second, progress and performance monitoring help contract managers to ensure that the contract is completed on time and within budget. By tracking the progress of the contractor's work and assessing their performance, contract managers can identify any areas where the project is at risk of falling behind schedule or exceeding budget. They can then take steps to address these issues and keep the project on track.

Third, progress and performance monitoring helps contract managers to improve the quality of the contractor's work. By providing feedback to the contractor on their performance, contract managers can help to identify and address any areas where the contractor's work is not meeting the required standards. This can lead to a better overall outcome for the project. Finally, progress and performance monitoring helps contract managers to build trust and rapport with the contractor. By working closely with the contractor to monitor progress and performance, contract managers can show the contractor that they are committed to the success of the project. This can help to create a more positive working environment and can lead to a more successful outcome for the project.

How to Implement Progress and Performance Monitoring

There are a number of different ways to implement progress and performance monitoring in contract management. Some of the most common methods include:

Project schedules: Project schedules are a valuable tool for tracking the progress of the contractor's work and identifying any potential delays. Project schedules should be developed in collaboration with the contractor and should be regularly updated to reflect the actual progress of the work.

Key performance indicators (KPIs): KPIs are measurable metrics that can be used to assess the contractor's performance against specific goals or objectives. KPIs should be carefully selected to ensure that they are relevant to the contract and that they provide meaningful insights into the contractor's performance.

Progress reports: Progress reports are a good way for the contractor to provide regular updates to the contract manager on the progress of the work. Progress reports should be clear and concise, and they should identify any challenges or issues that the contractor is facing.

Site visits and inspections: Site visits and inspections allow the contract manager to assess the contractor's progress and performance firsthand. Site visits and inspections should be conducted on a regular basis, and they should be focused on key areas of risk or concern.

Risk registers: Risk registers are used to identify and assess potential risks to the project. Risk registers should be regularly updated to reflect any changes in the risk profile of the project.

Analyzing Progress and Performance Monitoring Data

Once the contract manager has collected data on progress and performance, they need to be able to analyze this data to identify trends and patterns. This can be done using a variety of tools and techniques, such as data visualization tools, trend analysis, and root cause analysis. By effectively analyzing progress and performance monitoring data, contract managers can identify areas where the project is at risk of falling behind schedule, exceeding budget, or failing to meet the required quality standards. They can then take steps to address these issues and keep the project on track.

Progress and performance monitoring are essential components of effective contract management. By effectively monitoring progress and performance, contract managers can identify and address any potential problems early on, and ensure that the contract is completed on time, within budget, and to the required quality standards. Effective progress and performance monitoring requires the use of a variety of tools and techniques, as well as the ability to analyze data and make informed decisions.

CHAPTER 5: COMMUNICATION AND STAKEHOLDER MANAGEMENT

Communication and stakeholder management are essential components of effective contract management. They enable contract managers to build and maintain trust with stakeholders, identify and address potential problems early on, ensure that everyone is aligned on the goals and objectives of the contract, make informed decisions, and resolve disputes quickly and effectively.

5.1 Effective Communication Strategies

Communication is the process of exchanging information between two or more parties. It is essential for any relationship, but it is especially important in the context of contract management. Contract managers need to be able to communicate effectively with a variety of stakeholders, including the contractor, the client, and other stakeholders such as legal counsel and financial analysts.

Effective communication can help contract managers to build and maintain trust with stakeholders. Trust is essential for any successful relationship, and it is especially important in contract management. When stakeholders trust the contract manager, they are more likely to cooperate and be supportive. Contract managers can build trust by being honest and transparent in their communication, and by being responsive to stakeholders' needs. Effective communication can also help to identify and address potential problems early on. Communication can help contract managers to identify and address potential problems early on. This is important because it can help to prevent small problems from growing into bigger problems. Contract managers can identify potential problems by communicating regularly with stakeholders and by being proactive in seeking out feedback.

It can ensure that everyone is aligned on the goals and objectives of the contract. It is important for all stakeholders to be aligned on the goals and objectives of the contract. This will help to ensure that the contract is successful. Contract managers can ensure that everyone is aligned by communicating the goals and objectives of the contract clearly and concisely, and by seeking feedback from stakeholders.

Contract managers need to make a lot of decisions throughout the course of a contract. Effective communication can help them to make informed decisions by providing them with all of the necessary information. Contract managers can improve their decision-making by communicating regularly with stakeholders and by seeking their input. Disputes are inevitable in any contract. Effective communication can help contract managers to resolve disputes quickly and effectively. Contract managers can improve their dispute resolution skills by being clear and concise in their communication, and by being willing to listen to the other party's perspective.

These are some specific factors essential for effective communication in contract management:

Active voice is more concise and easier to read. For example, instead of saying "The contract was negotiated by the contract manager," say "The contract manager negotiated the contract." Additionally, positive language instead of negative language should be used. Positive language creates a more positive tone and makes your communication more effective. For example, instead of saying "We cannot accept your request," say "We appreciate your request, but we are unable to accept it at this time."

Specific examples should be used instead of generalizations. Specific examples make your communication more clear and persuasive. For example, instead of saying "The contractor is not meeting our expectations," say "The contractor has missed two deadlines and has not yet provided us with a revised schedule." It is also important to proofread communication before sending it. Proofreading helps to ensure that there are no errors in grammar or spelling. Errors in grammar and spelling can make your communication unprofessional and can lead to misunderstandings. The following are specific examples of how contract managers can use effective communication strategies in different situations:

When negotiating a contract: When negotiating a contract, contract managers need to be able to communicate effectively with the contractor in order to reach an agreement that is fair and equitable to both parties. This means being clear about the client's requirements and expectations, and being able to listen to and understand the contractor's concerns. Contract managers can also use effective communication strategies to build rapport with the contractor and to create a positive working environment.

When managing a contract: Once the contract is in place, contract managers need to be able to communicate effectively with all stakeholders in order to manage the contract effectively.

This means providing regular updates on the progress of the contract, addressing any concerns that stakeholders may have, and seeking feedback from stakeholders. Contract managers can also use effective communication strategies to build trust with stakeholders and to ensure that everyone is aligned on the goals and objectives of the contract.

When resolving disputes: If a dispute arises under the contract, contract managers need to be able to communicate effectively with all parties involved in order to resolve the dispute quickly and effectively. This means listening to all sides of the story and being able to identify a solution that is acceptable to everyone involved. Contract managers can also use effective communication strategies to build trust with the parties involved and to create a positive environment for resolution.

5.2 Stakeholder Engagement and Collaboration

Stakeholder engagement and collaboration are essential components of effective contract management. Stakeholders are individuals or groups who have an interest in the outcome of a contract. They may be directly involved in the contract, such as the contractor and the client, or they may be indirectly affected by the contract, such as employees, suppliers, and the community. Effective stakeholder engagement and collaboration can help contract managers identify and understand the needs and interests of stakeholders. By understanding the needs and interests of stakeholders, contract managers can develop contracts that are fair and equitable to all parties involved. This can help to prevent disputes and improve the overall success of the contract.

Effective stakeholder engagement and collaboration can also help contract managers identify and reduce the risk of disputes. When stakeholders are engaged in the contract process and feel that their needs are being considered, they are more likely to be supportive of the contract and less likely to raise disputes later on. Additionally, by working together to resolve any issues that may arise, contract managers and stakeholders can help to reduce the risk of disputes escalating into costly and time-consuming legal battles.

By collaborating with stakeholders, contract managers can identify and address potential problems early on, and ensure that the contract meets the needs of all parties involved. This can lead to a higher quality contract that is more likely to be successful. Furthermore, trust and rapport can be fostered with stakeholder engagement. By engaging with stakeholders in a meaningful way, contract managers can build trust and rapport with them. This can lead to a more positive and productive working relationship, which can benefit all parties involved.

The following strategies can help contract managers build effective stakeholder engagement and collaboration in contract management:

- Identify all relevant stakeholders. The first step in effective stakeholder engagement is to identify all relevant stakeholders. This includes individuals and groups who have an interest in the outcome of the contract, as well as those who may be directly or indirectly affected by the contract. This can be done by brainstorming, reviewing documentation, and conducting interviews.

- Understand the needs and interests of stakeholders. Once the stakeholders have been identified, it is important to understand their needs and interests. This can be done through interviews, surveys, and workshops. It is important to get a clear understanding of each stakeholder's perspective, as well as their concerns and expectations.

- Develop a stakeholder engagement plan. A stakeholder engagement plan should outline how the contract manager will engage with stakeholders throughout the contract lifecycle. The plan should identify the stakeholders, the communication channels that will be used, and the frequency of communication. The plan should also outline specific strategies for addressing the needs and interests of each stakeholder group.

- Communicate regularly and openly. It is important to communicate regularly and openly with stakeholders throughout the contract lifecycle. This means providing updates on the progress of the contract, addressing any concerns that stakeholders may have, and seeking feedback from stakeholders. Communication should be clear, concise, and timely.

- Collaborate with stakeholders. Contract managers should collaborate with stakeholders throughout the contract lifecycle. This means working with stakeholders to develop the contract, to manage the contract, and to resolve any disputes that may arise. Collaboration can take many different forms, such as meetings, workshops, and online collaboration tools.

Examples of how contract managers can use stakeholder engagement and collaboration strategies in different situations

When developing a contract: Contract managers can engage stakeholders in the contract development process by holding meetings, conducting surveys, and circulating draft contracts for feedback. This can help to ensure that the contract meets the needs of all parties involved and that everyone is aligned on the goals and objectives of the contract.

When managing a contract: Contract managers can communicate regularly with stakeholders by providing status updates, sending out progress reports, and holding regular meetings. This can help to keep everyone informed of the progress of the contract and to identify and address any potential problems early on. Contract managers can also collaborate with stakeholders by working together to resolve any issues that may arise, such as delays, budget overruns, or quality issues.

When resolving disputes: If a dispute arises under the contract, contract managers can collaborate with the parties involved to resolve the dispute quickly and effectively. This may involve mediation, arbitration, or litigation. Contract managers can also play a role in preventing disputes by engaging with stakeholders early on and by working to ensure that everyone's needs are being met.

5.3 Conflict Resolution and Dispute Avoidance

Conflict resolution and dispute avoidance are essential components of effective contract management. Contracts are complex agreements, and it is inevitable that disputes will arise from time to time. However, by taking steps to resolve conflicts quickly and effectively, and by taking steps to avoid disputes in the first place, contract managers can minimize the impact of disputes on the contract and the relationship between the parties involved.

Conflict Resolution

When a dispute arises under a contract, the contract manager should take a proactive approach to resolving it. The first step is to identify the source of the conflict. This may involve meeting with the parties involved to discuss their concerns and perspectives. It is important to listen carefully to both sides and to try to understand their positions. Once the source of the conflict has been identified, the contract manager can develop a plan to resolve it. This may involve mediation, arbitration, or litigation. The contract manager should choose the dispute resolution method that is most appropriate for the situation and that is acceptable to both parties. Once the plan has been developed, the contract manager should implement it and monitor the results. This may involve meeting with the parties involved to ensure that they are following the plan and that the conflict is being resolved. The contract manager should also be prepared to make adjustments to the plan as needed.

Dispute Avoidance

In addition to resolving conflicts, it is also important for contract managers to take steps to avoid disputes in the first place. Contracts should be clear and concise, and they should clearly outline the rights and obligations of each party. Ambiguity in a contract can lead to misunderstandings and disputes. The contract manager should also make sure that the contract is well-organized and easy to read. Contract managers should communicate regularly and openly with all parties involved in the contract. This can help to identify and address any potential problems early on before they escalate into disputes. Contract managers should also be responsive to the needs of all parties and should be willing to listen to their concerns. They should also be proactive and responsive to the needs of all parties involved in the contract. This means being willing to listen to their concerns and work with them to resolve any issues that may arise. It is also essential to be proactive in identifying and addressing potential problems early on.

Conflict resolution and dispute avoidance strategies can be used to negotiate a contract that is fair and equitable to all parties involved. For example, the contract manager can work with the contractor to develop a clear and concise contract that outlines the rights and obligations of each party. The contract manager can also discuss dispute resolution mechanisms with the contractor and include them in the contract.

Conflict resolution and dispute avoidance strategies Can also be used to manage conflicts that arise under the contract. For example, if a dispute arises over the quality of the contractor's work, the contract manager can meet with the contractor to discuss the issue and develop a plan to resolve it. The contract manager can also use mediation or arbitration to resolve the dispute.

CHAPTER 6:
BUDGETING AND
COST CONTROL

Budgeting and cost control are essential for contract managers to ensure that projects are completed within budget and on time. Contract managers need to create a budget for each contract and then track actual costs against the budget to identify and address any potential overruns.

6.1 Budget Development
and Management

Budget development and management are essential components of effective contract management. Contract managers are responsible for ensuring that contracts are executed within budget and on time. This requires them to have a deep understanding of the contract requirements, the costs associated with delivering the contract, and the risks that could impact the budget.

Budget Development

The first step in budget development is to understand the scope of work and the requirements of the contract. This includes identifying all of the goods and services that need to be procured, as well as the associated costs. Once the scope of work is understood, the contract manager can develop a budget that allocates the necessary resources to each task. The budget should be realistic and achievable, and it should be based on historical data and industry benchmarks. It is also important to include a contingency fund in the budget to cover unexpected expenses. Once the budget is developed, it should be reviewed and approved by the appropriate stakeholders. This may include the client, the project manager, and the financial team.

Budget Management

Once the budget is approved, the contract manager is responsible for tracking actual costs against the budget. This can be done through a variety of methods, such as timesheets, expense reports, and invoices. If the contract manager identifies any potential overruns, they need to take corrective action immediately. This may involve renegotiating the contract with the client, finding alternative suppliers, or reducing the scope of work. The contract manager should also regularly report on the budget status to the stakeholders. This helps to keep everyone informed of the project's financial progress and to identify any potential problems early on.

Best Practices for Budget Development and Management

Involve stakeholders early on. It is important to involve all relevant stakeholders in the budget development process. This will help to ensure that the budget is realistic and achievable, and that it meets the needs of all parties involved.

Use historical data and industry benchmarks. When developing the budget, it is important to use historical data and industry benchmarks to estimate costs. This will help to ensure that the budget is realistic and achievable.

Include a contingency fund. It is important to include a contingency fund in the budget to cover unexpected expenses. This will help to prevent the project from going over budget if there are any unforeseen costs.

Track actual costs against the budget regularly. The contract manager should track actual costs against the budget regularly. This will help to identify any potential overruns early on and to take corrective action.

Report on the budget status to stakeholders regularly. The contract manager should regularly report on the budget status to the stakeholders. This will help to keep everyone informed of the project's financial progress and to identify any potential problems early on.

Challenges of Budget Development and Management in Contract Management

Uncertainty: The scope of work and the requirements of a contract can often change over time. This can make it difficult to develop a realistic and accurate budget.

Complexity: Contracts can be complex and involve a variety of different stakeholders. This can make it difficult to track actual costs against the budget and to identify potential overruns.

Pressure to deliver on time and within budget: Contract managers are often under pressure to deliver projects on time and within budget. This can make it difficult to take corrective action if there are any potential overruns.

Overcoming the Challenges of Budget Development and

Management in Contract Management

Involve stakeholders early on. Involving all relevant stakeholders early on in the budget development process can help to reduce uncertainty and ensure that the budget is realistic and achievable. It is important to use project management tools and techniques. There are a number of project management tools and techniques that can be used to track actual costs against the budget and to identify potential overruns. Contract managers should be proactive in identifying and addressing potential overruns. If there are any changes to the scope of work or the requirements of the contract, the budget should be updated accordingly. Finally, they should communicate regularly with stakeholders about the budget status. This will help to keep everyone informed of the project's financial progress and to build trust and rapport.

6.2 Cost Estimation and Control Techniques

Cost estimation and control are essential components of effective contract management. Contract managers are responsible for ensuring that contracts are executed within budget and on time. This requires them to have a deep understanding of the costs associated with delivering the contract, the risks that could impact the budget, and the techniques that can be used to control costs.

Cost Estimation

Cost estimation is the process of forecasting the costs associated with delivering a contract. It is an important step in the contract management process, as it allows contract managers to set realistic budgets and identify and mitigate potential cost overruns. There are a number of different cost estimation techniques that can be used by contract managers. The most appropriate technique will depend on the nature of the contract, the complexity of the work, and the availability of data. Some common cost estimation techniques include:

Analogous estimating: This technique involves using the costs of similar projects to estimate the costs of the current project.

Parametric estimating: This technique uses historical data to develop relationships between different cost drivers and the total cost of a project.

Bottom-up estimating: This technique involves estimating the costs of each individual task in the project and then summing them together to get an overall cost estimate.

Cost Control

Cost control is the process of monitoring and managing costs to ensure that they stay within budget. Contract managers should use a variety of cost control techniques to identify and address potential cost overruns early on. Some common cost control techniques include:

Earned value management (EVM): EVM is a project management technique that can be used to track the progress of a project and to identify potential cost overruns. EVM compares the planned value of the work completed to the actual cost of the work completed. If there is a difference between the planned value and the actual cost, this indicates that the project is over or under budget.

Variance analysis: Variance analysis is a technique that can be used to identify the causes of cost overruns. Variance analysis compares the actual cost of a project to the budget and to the costs of similar projects. This can help contract managers to identify areas where costs are higher than expected.

Risk management: Risk management is the process of

identifying, assessing, and mitigating risks. Contract managers should use a risk management process to identify and mitigate potential cost risks. For example, contract managers may include a contingency fund in the budget to cover unexpected costs.

Contract managers use a variety of cost estimation and control techniques to ensure that contracts are executed within budget and on time. They also understand the importance of communicating with stakeholders about the budget and of taking corrective action early on if there are any potential cost overruns. These are some of the guides for contract managers on how to use cost estimation and control techniques effectively:

Use the right cost estimation technique: The most appropriate cost estimation technique will depend on the nature of the contract, the complexity of the work, and the availability of data. Contract managers should choose the technique that is most likely to provide an accurate cost estimate.

Get input from stakeholders: Contract managers should get input from all relevant stakeholders when estimating costs. This will help to ensure that the cost estimate is realistic and that it meets the needs of all parties involved.

Monitor costs regularly: Contract managers should monitor costs regularly throughout the contract lifecycle. This will help to identify potential cost overruns early on and to take corrective action.

Communicate with stakeholders: Communication cannot be overemphasized. Contract managers should communicate regularly with stakeholders about the budget and about any potential cost overruns. This will help to build trust and rapport with stakeholders and to ensure that everyone is working towards the same goal.

6.3 Tracking and Managing

Project Expenses

Tracking and managing project expenses is essential for effective contract management. Contract managers are responsible for ensuring that contracts are executed within budget and on time. This requires them to have a clear understanding of the project's expenses, to track those expenses against the budget, and to identify and take corrective action if there are any potential overruns. There are a number of different ways to track and manage project expenses. The most appropriate method will depend on the size and complexity of the project, as well as the contract manager's personal preferences. Some common methods for tracking and managing project expenses include:

Timesheets: Timesheets can be used to track the time spent by employees on different project tasks. This information can then be used to calculate the labor costs associated with the project.

Expense reports: Expense reports can be used to track the costs of goods and services incurred on behalf of the project. This may include items such as travel expenses, office supplies, and equipment rental.

Project management software: There are a number of project management software programs that can help contract managers track project expenses. These programs typically allow users to create budgets, track actual costs, and generate reports.

Contract managers use a variety of methods to track and manage project expenses. They also understand the importance of communicating with stakeholders about the budget and of taking corrective action early on if there are any potential cost overruns. The following provides a guide on how to track and manage project expenses effectively.

Develop a budget breakdown. Before starting the project, contract managers should develop a detailed budget breakdown. This should include all of the anticipated costs associated with the project, such as labor costs, material costs, and travel expenses.

Track expenses regularly. Contract managers should track expenses regularly throughout the project lifecycle. This will help to identify potential cost overruns early on and to take corrective action.

Use project management software. Project management software can help contract managers track project expenses and identify potential cost overruns.

Communicate with stakeholders. Contract managers should communicate regularly with stakeholders about the budget and about any potential cost overruns. This will help to build trust and rapport with stakeholders and to ensure that everyone is working towards the same goal.

CHAPTER 7: LEGAL AND REGULATORY COMPLIANCE

L egal and regulatory compliance is essential for contract managers to ensure that their contracts comply with all applicable laws and regulations. This includes laws and regulations related to contract formation, performance, and termination. Contract managers need to be aware of all relevant laws and regulations and ensure that their contracts are drafted and executed in accordance with them.

7.1 Understanding Legal Obligations

Effective contract managers have a deep understanding of the legal obligations that they and their organizations have under the contracts that they manage. This understanding is essential for ensuring that contracts are performed in accordance with the law and that the organization's interests are protected.

Common Legal Obligations in Contracts

There are a number of common legal obligations that parties to contracts have. These obligations include:

- An obligation to perform the contract in good faith. This means that the parties must act honestly and fairly in their dealings with each other.

- An obligation to comply with the terms of the contract. This means that the parties must do what they agreed to do in the contract.
- An obligation to avoid causing harm to the other party. This means that the parties must take reasonable care to avoid causing each other losses or damages.

In addition to these common legal obligations, contracts may also contain specific legal obligations that are tailored to the specific nature of the contract. For example, a construction contract may contain specific obligations regarding the quality of the work to be performed and the materials to be used.

Understanding the Law

Contract managers need to have a basic understanding of the law in order to identify and understand the legal obligations that they and their organizations have under the contracts that they manage. This includes understanding the law of contract, the law of tort, and the law of property. Contract managers may also need to have a more specific understanding of the law depending on the type of contracts that they manage. For example, contract managers who manage government contracts need to have a good understanding of the public procurement laws and regulations that apply to their contracts.

Identifying Legal Obligations
in Contracts

Contract managers can identify the legal obligations that they and their organizations have under the contracts that they manage by carefully reading and reviewing the contracts. Contract managers should also pay attention to any specific laws and regulations that may apply to their contracts. If contract managers are unsure about the meaning of a particular legal obligation, they should consult with an attorney.

Managing Legal Obligations

Once legal obligations for the contracts that they manage have been identified, they need to take steps to manage those obligations. They must ensure that the organization complies with its obligations. Steps must be taken to ensure that the organization complies with all of its legal obligations under its contracts. This may involve developing policies and procedures, training staff, and monitoring compliance.

Contract managers need to identify and mitigate any risks associated with the organization's legal obligations under its contracts. This may involve purchasing insurance, obtaining indemnities from third parties, and negotiating favorable contract terms. They also need to be prepared to respond to disputes that may arise under the contracts that they manage. This may involve negotiating with the other party, engaging in mediation or arbitration, or litigating the dispute in court.

7.2 Environmental and Safety Regulations

Environmental and safety regulations are essential for protecting the environment and the safety of workers and the public. Organizations that fail to comply with these regulations can face significant financial and legal penalties, and may also damage their reputation.

Contract managers play a critical role in ensuring that their organizations comply with environmental and safety regulations. They are responsible for developing and implementing environmental and safety management plans, monitoring and reporting on environmental and safety performance, and taking corrective action when necessary.

To effectively manage environmental and safety obligations, contract managers should:

Identify and understand the environmental and safety regulations that apply to their organization's contracts. This can be done by reviewing the contracts themselves, as well as by consulting with environmental and safety professionals. They should also develop and implement environmental and safety management plans (ESMPs). ESMPs help to identify, assess, and manage the environmental and safety risks associated with a particular project or contract.

It is also important to monitor and report on environmental and safety performance. This may involve tracking emissions, waste generation, and accidents. Monitoring and reporting data can be used to identify areas where the organization is at risk of non-compliance with environmental and safety regulations. Taking corrective action when necessary is critical. If contract managers identify any environmental or safety problems, they need to take corrective action immediately. This may involve changing work practices, implementing new controls, or training staff.
Contract managers should also work closely with the other party to the contract to ensure that they are taking appropriate corrective action.

7.3 Compliance Monitoring and Reporting

Compliance monitoring and reporting are essential components of effective contract management. Organizations of all sizes and industries are subject to a variety of compliance obligations, and contract managers play a critical role in ensuring that contracts are compliant with all applicable laws, regulations, industry standards, and contractual agreements.

Compliance monitoring involves identifying and tracking compliance risks, as well as assessing the effectiveness of compliance controls. Compliance reporting involves communicating the results of compliance monitoring to stakeholders. Contract managers can implement a compliance monitoring and reporting program by following these processes:

Identify the relevant compliance obligations. This may involve reviewing the contracts themselves, as well as conducting a risk assessment.

Develop compliance controls. Once the compliance obligations have been identified, contract managers need to develop and implement controls to mitigate the risks. This may involve developing standard operating procedures, training staff, and conducting audits.

Monitor compliance performance. Contract managers need to monitor compliance performance on an ongoing basis to ensure that the controls implemented are effective. This may involve tracking compliance metrics, conducting audits, and investigating compliance incidents.

Report on compliance performance. Contract managers need to communicate the results of compliance monitoring to stakeholders. This may involve reporting on compliance metrics, audit findings, and corrective actions taken.

Compliance monitoring and reporting can be a complex and time-consuming task, but it is essential for organizations that want to avoid the risks associated with non-compliance. By following the steps outlined above, contract managers can help their organizations to maintain a high level of compliance and to protect themselves from legal and financial penalties. These are additional factors for contract managers to consider in compliance monitoring and reporting:

Keep up with changes in the law and regulations. Compliance laws and regulations are constantly changing, so it is important for contract managers to keep up with the latest changes. This can be done by reading industry publications, attending conferences, and networking with other contract managers.

Seek expert advice when needed. Contract managers should not hesitate to seek expert advice from legal counsel or compliance officers if they have any questions or concerns about compliance obligations. This is especially important for contracts that involve complex or sensitive compliance issues.

Document everything. Contract managers should document everything related to their compliance monitoring and reporting activities, including all communications with stakeholders, audit reports, and corrective actions taken. This documentation can be invaluable if a compliance issue does arise.

In addition to the above, contract managers can also implement a number of other measures to ensure effective compliance monitoring and reporting. These include:

Using technology to automate tasks. There are a number of software solutions available that can help contract managers to automate compliance monitoring and reporting tasks. This can free up time for contract managers to focus on other important tasks, such as negotiating and managing contracts.

Developing a compliance culture. Contract managers can help to develop a compliance culture within their organization by promoting awareness of compliance obligations and by providing training to staff on compliance procedures. This can help to create an environment where everyone is responsible for compliance and where compliance risks are more likely to be identified and addressed.

Working closely with other stakeholders. Contract managers

need to work closely with other stakeholders, such as legal counsel, compliance officers, and internal auditors, to ensure that compliance obligations are being met. This may involve sharing information, collaborating on compliance initiatives, and participating in joint audits.

CHAPTER 8: RISK MANAGEMENT

Risk management is the process of identifying, assessing, and mitigating risks associated with contracts. Contract managers play a vital role in risk management by identifying and assessing potential risks, developing and implementing risk mitigation strategies, and monitoring the effectiveness of these strategies.

8.1 Identifying and Assessing Project Risks

Identifying project risks is the process of recognizing potential events or conditions that could adversely affect a project's objectives. Assessing project risks is the process of analyzing the likelihood and impact of identified risks in order to prioritize them and develop mitigation strategies. Contract managers play a critical role in identifying and assessing project risks. They have a deep understanding of the contracts that they manage, as well as the potential risks associated with those contracts. By proactively identifying and assessing project risks, contract managers can help their organizations avoid costly delays, disruptions, and disputes.

There are a number of different methods that can be used

to identify and assess project risks. One common approach is to conduct a risk assessment workshop. This involves bringing together a group of stakeholders with different perspectives on the project to identify potential risks. The workshop participants can then use a variety of tools and techniques to assess the likelihood and impact of each risk. Another common approach to identifying and assessing project risks is to use a risk register. A risk register is a document that lists all of the identified risks associated with a project, as well as their likelihood, impact, and mitigation strategies. The risk register can be updated on an ongoing basis as new risks are identified or as the project's environment changes.

Regardless of the method used, it is important to involve all relevant stakeholders in the risk identification and assessment process. This includes the contract manager, the project manager, the client, and any other subcontractors or suppliers who will be involved in the project. Once the project risks have been identified and assessed, the contract manager can develop mitigation strategies. Mitigation strategies are designed to reduce the likelihood or impact of project risks. Some common mitigation strategies include:

Avoiding the risk: This involves taking steps to avoid the risk from occurring in the first place. For example, if a contract manager is concerned about the risk of a supplier failing to deliver on time, they may choose to work with a different supplier.
Transferring the risk: This involves transferring the risk to another party. For example, a contract manager may purchase insurance to mitigate the risk of financial losses due to a delay in the project.

Reducing the impact of the risk: This involves taking steps to reduce the impact of a risk if it does occur. For example, a contract manager may develop a contingency plan in the event that a key supplier fails to deliver on time.

By proactively identifying and assessing project risks, and by developing effective mitigation strategies, contract managers can help their organizations avoid costly delays, disruptions, and disputes. The following are additional factors for effective contract management in identifying and assessing project risks.

Keep up to date with industry trends and best practices. Contract managers should regularly review industry trends and best practices for risk identification and assessment. This will help them to ensure that they are using the most effective methods and tools.

Seek expert advice when needed. If contract managers are unsure about how to identify or assess a particular risk, they should seek expert advice from a risk management professional.

Document everything. Contract managers should document all aspects of their risk identification and assessment process. This includes the risks that were identified, the assessment results, and the mitigation strategies that were developed. This documentation can be invaluable if a dispute does arise.

8.2 Risk Mitigation and Contingency Planning

Risk mitigation is the process of taking steps to reduce the likelihood or impact of risks. Contingency planning is the process of developing plans to respond to risks if they do occur. Contract managers play a critical role in risk mitigation and contingency planning. They have a deep understanding of the contracts that they manage, as well as the potential risks associated with those contracts. By proactively mitigating risks and developing contingency plans, contract managers can help their organizations to avoid costly delays, disruptions, and disputes.

There are a number of different risk mitigation strategies that can be used. Some common risk mitigation strategies include:

Avoiding the risk: This involves taking steps to avoid the risk from occurring in the first place. For example, if a contract manager is concerned about the risk of a supplier failing to deliver on time, they may choose to work with a different supplier.

Transferring the risk: This involves transferring the risk to another party. For example, a contract manager may purchase insurance to mitigate the risk of financial losses due to a delay in the project.

Reducing the impact of the risk: This involves taking steps to reduce the impact of a risk if it does occur. For example, a contract manager may develop a contingency plan in the event that a key supplier fails to deliver on time.

Contingency plans should be developed for all significant risks associated with a contract. Contingency plans should outline the steps that will be taken to respond to a risk if it does occur. Contingency plans should be regularly reviewed and updated to ensure that they are still effective. When developing contingency plans, it is important to consider the likelihood of the risk occurring, the potential impact of the risk if it does occur, the resources that are available to respond to the risk, and the time constraints involved in responding to the risk.

Contract managers will work closely with other stakeholders, such as the project manager and the client, to develop and implement risk mitigation and contingency plans. By working together, these stakeholders can develop a comprehensive plan to manage the risks associated with the contract. The following are examples of how contract managers can use risk mitigation and contingency planning to avoid costly delays, disruptions, and disputes.

Example 1: A contract manager was concerned about the risk of a supplier failing to deliver on time. To mitigate this risk, the contract manager negotiated a liquidated damages clause into the contract. This clause would require the supplier to pay the contract manager a certain amount of money for each day that the delivery was late. This clause helped to motivate the supplier to deliver on time.

Example 2: A contract manager was concerned about the risk of a key employee leaving the company during the project. To mitigate this risk, the contract manager developed a contingency plan that identified other employees who could step in if the key employee left. This helped to ensure that the project would continue to progress even if the key employee left.

Example 3: A contract manager was concerned about the risk of a natural disaster disrupting the project. To mitigate this risk, the contract manager purchased insurance that would cover the cost of any delays or damages caused by a natural disaster. This insurance helped to protect the organization from financial losses if a natural disaster did occur.

These are examples of how contract managers have used risk mitigation and contingency planning to avoid costly delays, disruptions, and disputes. By proactively managing risks, contract managers can help their organizations achieve their goals and objectives.

8.3 Insurance and Bonding in Construction Contracts

Construction contracts are complex and involve a significant amount of risk. Construction contract insurance and bonding can help to mitigate these risks and protect the parties involved in the contract. Construction contract insurance is a type of insurance that protects the parties involved in a construction contract from financial losses due to events such as property damage, personal injury, and lawsuits. There are a number of different types of construction contract insurance, including:

Builders risk insurance: This type of insurance covers property damage and loss during the course of construction.

Commercial general liability insurance: This type of insurance covers personal injury and property damage caused by the contractor or its subcontractors.

Professional liability insurance: This type of insurance covers errors and omissions by the contractor or its subcontractors.

Construction contract bonding is a type of financial guarantee that ensures that the contractor will complete the project according to the terms of the contract. There are three main types of construction contract bonds:

Bid bonds: These bonds guarantee that the contractor will submit a bid in good faith and will enter into a contract if awarded the bid.

Performance bonds: These bonds guarantee that the contractor will complete the project according to the terms of the contract.

Payment bonds: These bonds guarantee that the contractor will pay its subcontractors and suppliers.

Contract managers play a critical role in ensuring that construction contracts are properly insured and bonded. They have a deep understanding of the risks associated with construction projects and can work with the client and the contractor to develop an insurance and bonding program that meets the needs of all parties involved. The following are the key tasks that effective contract managers perform in the context of construction contract insurance and bonding.

Identifying the risks associated with the project. Contract managers need to identify all of the potential risks associated with the construction project. This includes risks such as property damage, personal injury, lawsuits, and financial losses due to delays or incomplete work.

Assessing the likelihood and impact of each risk. Once the risks have been identified, there is a need to assess the likelihood and impact of each risk. This will help them to determine the appropriate level of insurance and bonding coverage.

Developing an insurance and bonding program. Contract managers need to develop an insurance and bonding program that meets the needs of all parties involved. This program should be tailored to the specific risks associated with the project.

Contract managers need to work with the client and the contractor to implement the insurance and bonding program. This includes negotiating the terms of the insurance and bonding policies and ensuring that the contractor has obtained the required coverage. Finally, they need to monitor the insurance and bonding program throughout the course of the project. This includes ensuring that the contractor maintains the required coverage and that any claims are reported and processed promptly.

CHAPTER 9: PROJECT CLOSEOUT AND EVALUATION

Project closeout and evaluation are the final stages of the contract management process. Contract managers are responsible for ensuring that all project deliverables have been met, that all contractual obligations have been fulfilled, and that the project is closed out in a timely and orderly manner. Contract managers also need to evaluate the project to identify lessons learned and to improve future project performance.

9.1 Final Inspections and Punch Lists

Final inspections and punch lists are essential parts of any construction project. Final inspections are conducted to ensure that the project has been completed in accordance with the terms of the contract, while punch lists are used to document any items that need to be completed or corrected before the project can be considered complete.

Contract managers play a critical role in ensuring that final inspections and punch lists are conducted properly. They have a deep understanding of the contract and the project, and they can work with the contractor and the client to ensure that all of the necessary items are addressed. The following are some of the key tasks that effective contract managers perform in the context of final inspections and punch lists.

Schedule the final inspection: Contract managers need to schedule the final inspection with the contractor and the client. This should be done well in advance of the expected completion date of the project.

Coordinate the final inspection: Contract managers need to coordinate the final inspection with all of the relevant stakeholders. This includes the contractor, the client, the architect, and any other subcontractors or suppliers who were involved in the project.

Prepare the punch list: Contract managers need to prepare a punch list of any items that need to be completed or corrected before the project can be considered complete. The punch list should be detailed and specific, and it should include a deadline for each item.

Follow up on the punch list: Contract managers need to follow up with the contractor to ensure that all of the items on the punch list have been completed or corrected. This may involve visiting the site regularly and inspecting the work.

Finalize the project: Once all of the items on the punch list have been completed or corrected, the contract manager can finalize the project. This may involve signing a certificate of completion and making the final payment to the contractor. The are some examples of how contract managers can use final inspections and punch lists to avoid costly delays, disruptions, and disputes:

Example 1: A contract manager was concerned that the contractor would not complete the project on time. To avoid this, the contract manager scheduled a final inspection for a date that was earlier than the expected completion date of the project. This gave the contractor more time to complete the work and to correct any deficiencies.

Example 2: A contract manager was concerned about the quality of the contractor's work. To avoid this, the contract manager involved the architect and other subcontractors in the final inspection. This helped to ensure that all of the work was completed to the satisfaction of all parties involved.

Example 3: A contract manager was concerned about the contractor completing all of the items on the punch list. To avoid this, the contract manager followed up with the contractor regularly and inspected the work frequently. This helped to ensure that all of the items on the punch list were completed before the project was finalized.

These are a few examples of how contract managers can use final inspections and punch lists to avoid costly delays, disruptions, and disputes. By properly managing final inspections and punch lists, contract managers can help to ensure that construction projects are completed successfully.

9.2 Contractual Obligations and Deliverables

Contractual obligations are the legal responsibilities that the parties to a contract agree to undertake. Contractual deliverables are the specific products or services that the contractor is obligated to provide to the client under the terms of the contract.

Contract managers play a critical role in ensuring that contractual obligations and deliverables are met. They have a deep understanding of the contract and the project, and they can work with the contractor and the client to ensure that all of the necessary requirements are addressed. These are some of the key tasks that contract managers perform in the context of contractual obligations and deliverables.

Identify the contractual obligations and deliverables: Contract managers need to identify all of the contractual obligations and deliverables that are specified in the contract. This includes reviewing the contract carefully and identifying all of the specific requirements that the contractor is obligated to meet.

Develop a project plan: Contract managers need to develop a project plan that outlines the steps that will be taken to meet the contractual obligations and deliverables. This plan should include a timeline, budget, and resource allocation plan.

Monitor the project's progress: Contract managers need to monitor the project progress to ensure that the contractor is meeting the contractual obligations and deliverables. This may involve reviewing progress reports, attending project meetings, and inspecting the work.

Manage risks: Contract managers need to manage the risks associated with the project. This may involve identifying and assessing risks, developing mitigation strategies, and monitoring the effectiveness of those strategies.

Communicate with stakeholders: Contract managers need to communicate regularly with the contractor, the client, and other stakeholders. This communication should include updates on the project's progress, any issues that arise, and any changes that need to be made to the project plan. The following are examples of how contract managers can use their knowledge of contractual obligations and deliverables to avoid costly delays, disruptions, and disputes.

Example 1: A contract manager was concerned that the contractor was not meeting the contractual obligations and deliverables for a software development project. To address this issue, the contract manager met with the contractor to discuss the specific areas where they were falling behind. The contract manager then worked with the contractor to develop a plan to get the project back on track. This plan included additional resources and a revised timeline. As a result of the contract manager's intervention, the contractor was able to meet the contractual obligations and deliverables and the project was completed successfully.

Example 2: A contract manager was concerned about the potential for a dispute over the scope of work on a construction project. To avoid this, the contract manager worked with the client and the contractor to develop a detailed scope of work statement. The scope of work statement clearly defined the contractual obligations and deliverables for the project. By having a clear and concise scope of work statement, the contract manager was able to avoid any disputes over the scope of work.

Example 3: A contract manager was concerned about the risk of a delay in the delivery of a critical piece of equipment for a manufacturing project. To mitigate this risk, the contract manager worked with the supplier to develop a contingency plan. The contingency plan included alternative suppliers in case the primary supplier was unable to deliver the equipment on time. As a result of the contract manager's planning, the project was able to stay on schedule even when the primary supplier was delayed in delivering the equipment.

9.3 Lessons Learned and Post-Project Evaluation

Lessons learned and post-project evaluation are essential components of effective contract management. By taking the time to reflect on what went well and what could be improved, contract managers can help to ensure that future projects are more successful. Lessons learned are the insights and knowledge gained from the experience of a project. They can be identified at any time during the project, but it is especially important to conduct formal lessons learned exercises at the end of the project. This exercise should involve all of the stakeholders involved in the project, including the contract manager, the contractor, the client, and any other subcontractors or suppliers.

Post-project evaluation is a more comprehensive assessment of the project. It typically involves reviewing the project's goals and objectives, assessing the project's performance, and identifying areas for improvement. The post-project evaluation should be documented in a report that can be used to inform future project management practices.

Contract managers play a critical role in both lessons learned and post-project evaluation. They have a deep understanding of the project, and they can work with the stakeholders to identify and document lessons learned and to develop recommendations for improvement. The following are some of the key tasks that contract managers perform in the context of lessons learned and post-project evaluation.

Identify lessons learned: Contract managers need to identify lessons learned from the project. This can be done by interviewing stakeholders, reviewing project documentation, and analyzing project performance.

Document lessons learned: Contract managers need to document the lessons learned from the project. This documentation should be clear, concise, and actionable.

Share lessons learned with stakeholders: Contract managers need to share the lessons learned from the project with the stakeholders. This can be done through meetings, reports, or other communication channels.

Develop recommendations for improvement: Contract managers need to develop recommendations for improvement based on the lessons learned from the project. These recommendations should be specific, measurable, achievable, relevant, and time-bound.

Implement recommendations for improvement: Contract managers need to work with the stakeholders to implement the recommendations for improvement. This may involve developing new processes, procedures, or templates.

CHAPTER 10: TECHNOLOGY AND TOOLS IN CONTRACT MANAGEMENT

T echnology and tools play a vital role in helping contract managers to automate tasks, streamline workflows, and improve efficiency.

10.1 Construction Management Software

Construction management software (CMS) is a type of software that helps construction professionals to manage projects more effectively. CMS can help with a variety of tasks, including project planning and scheduling, cost management, document management, communication and collaboration, and risk management.

Contract managers can use CMS to improve their performance in a number of ways. For example, CMS can help contract managers in the following ways:

- Keep track of contractual obligations and deliverables.

CMS can be used to create a database of all of the contractual obligations and deliverables for a project. This can help contract managers to ensure that all of the requirements are met and that the project is completed on time and on budget.

- Manage risks. CMS can be used to identify, assess, and manage risks associated with a project. This can help contract managers to avoid costly delays and disruptions.

- Communicate with stakeholders. CMS can be used to communicate with stakeholders throughout the project lifecycle. This can help to keep everyone informed of the project's progress and to resolve any issues that arise.

- Generate reports. CMS can be used to generate reports on a variety of project metrics, such as cost, schedule, and risk. This information can be used to track the project's progress and to make informed decisions about the project.

The following are some specific examples of how contract managers can use CMS to improve their performance.

Example 1: A contract manager was concerned about the risk of a delay in the delivery of a critical piece of equipment for a construction project. To mitigate this risk, the contract manager used CMS to develop a contingency plan. The contingency plan identified alternative suppliers in case the primary supplier was unable to deliver the equipment on time. As a result of the contract manager's planning, the project was able to stay on schedule even when the primary supplier was delayed in delivering the equipment.

Example 2: A contract manager was responsible for managing a complex construction project with multiple stakeholders. To keep everyone informed of the project's progress and to resolve any issues that arose, the contract manager used CMS to create a communication portal. The communication portal allowed stakeholders to access project updates, share documents, and collaborate on tasks. This helped to improve communication and collaboration among the stakeholders and to ensure that the project was completed successfully.

Example 3: A contract manager was responsible for managing a construction project with a tight budget. To track the project's costs and make sure that the project was staying on budget, the contract manager used CMS to generate cost reports. The cost reports provided the contract manager with detailed information about the project's expenditures. This information allowed the contract manager to identify areas where costs could be reduced and to keep the project on budget.

10.2 Building Information Modeling

Building Information Modeling (BIM) is a process for creating and managing digital representations of physical and functional characteristics of a facility. BIM is used in the construction industry to improve the efficiency and effectiveness of project planning, design, construction, and operation.

Contract managers can use BIM to improve their performance in a number of ways. For example, BIM can help contract managers to better understand the project scope of work. BIM can be used to create a 3D model of the project that includes all of the project components. This can help contract managers to better understand the project scope of work and to identify any potential conflicts or interference.

BIM can also be used to share information with stakeholders throughout the project lifecycle. This can help to improve communication and collaboration among the stakeholders and to ensure that everyone is on the same page. Furthermore, BIM can be leveraged to reduce the risk of errors and omissions.

BIM has proved to be effective in identifying errors and omissions in the project design and coordinating the work of different contractors. This can help to reduce the risk of costly delays and disruptions. Improving quality and productivity can also benefit BIM technology. For example, BIM can be used to generate shop drawings and to prefabricate components. The following are three examples of how contract managers can use BIM to improve their performance.

Example 1: A contract manager was responsible for managing a complex construction project with multiple stakeholders and contractors. To improve communication and collaboration among the stakeholders, the contract manager used BIM to create a shared project model. The shared project model allowed stakeholders to access the project model, view the most up-to-date project information, and comment on the project design. This helped to improve communication and collaboration among the stakeholders and to ensure that everyone was on the same page.

Example 2: A contract manager was responsible for managing a construction project with a tight budget. To reduce the risk of errors and omissions in the project design, the contract manager used BIM to conduct clash detection. Clash detection is a process that uses BIM to identify potential conflicts or interference between different project components. By identifying and resolving potential conflicts and interference early in the project lifecycle, the contract manager was able to avoid costly delays and disruptions.

Example 3: A contract manager was responsible for managing a construction project for a LEED-certified building. To improve the quality and productivity of the construction process, the contract manager used BIM to generate shop drawings and to prefabricate components. Using BIM to generate shop drawings and to prefabricate components helped to improve the accuracy and efficiency of the construction process and to reduce the risk of errors and omissions.

10.3 Automation and Data Analytics

Automation and data analytics are two rapidly evolving technologies that are having a significant impact on the field of contract management. Contract managers can use these technologies to improve their performance in a number of ways. Automation can be used to streamline many of the repetitive and time-consuming tasks associated with contract management, such as contract creation and review, contract negotiation, and contract tracking. This can free up contract managers to focus on more strategic tasks, such as risk management and relationship management.

Data analytics can be used to extract valuable insights from contract data. This information can be used to improve the efficiency and effectiveness of the contract management process, as well as to make better business decisions. The following are examples of how contract managers can use automation and data analytics to improve their performance.

- Use automation to streamline contract creation and review. Contract automation software can be used to generate and review contracts quickly and accurately. This can help to reduce the risk of errors and omissions and to improve the efficiency of the contract management process.

- Use automation to streamline contract negotiation. Contract negotiation software can be used to automate the negotiation process and to track the progress of negotiations. This can help to reduce the time it takes to negotiate contracts and to improve the outcome of negotiations.

- Use automation to track contract performance. Contract tracking software can be used to track the performance of contracts and to identify any potential problems. This can help to ensure that contracts are being performed as expected and to mitigate risks.

- Use data analytics to identify trends and patterns. Contract data analytics can be used to identify trends and patterns in contract data. This information can be used to improve the efficiency of the contract management process and to make better business decisions.

- Use data analytics to assess risk. Contract data analytics can be used to assess the risk associated with contracts. This information can be used to develop and implement risk mitigation strategies.

Contract managers can consider the following additional factors in using automation and data analytics.

Start small: It is important to start small when implementing automation and data analytics in contract management. This will help you to avoid any major disruptions to your existing processes and to learn how to use these technologies effectively.

Get buy-in from stakeholders: It is important to get buy-in from stakeholders before implementing automation and data analytics in contract management. This will help to ensure that everyone is on the same page and that the new technologies are used effectively.

Invest in training: It is important to invest in training for your staff on how to use automation and data analytics software. This will help to ensure that your staff is able to use these technologies effectively to improve their performance.

Monitor the results: It is important to monitor the results of implementing automation and data analytics in contract management. This will help you to identify any areas where you can make improvements.

CONCLUSION

Contract management is a critical function for any organization. By effectively managing contracts, organizations can reduce risk, improve performance, and achieve their goals. This book has provided a comprehensive overview of contract management, from the basics of contract law to the latest trends in technology.

Throughout the book, we have emphasized the importance of effective contract management for organizational success. We have also provided practical advice and tools to help organizations improve their contract management practices. Contract management is a complex and challenging discipline, but it is essential for any organization that wants to succeed. By following the principles and practices outlined in this book, organizations can improve their contract management capabilities and achieve their goals more effectively. The field of contract management is constantly evolving, as new technologies and trends emerge. These are a few of the key trends that we expect to see in the coming years:

Increased use of automation and data analytics. Automation and data analytics are already having a significant impact on contract management, and their use is expected to continue to grow in the coming years. Automation can be used to streamline many of the repetitive and time-consuming tasks associated with contract management, such as contract creation and review, contract negotiation, and contract tracking. Data analytics can be used to extract valuable insights from contract data, which can be used to improve the efficiency and effectiveness of the contract management process, as well as to make better business decisions.

Greater focus on risk management. Risk management is becoming increasingly important in contract management, as organizations face a growing number of risks, such as cyber threats, supply chain disruptions, and regulatory changes. Contract managers will need to develop and implement robust risk management strategies to mitigate these risks and protect their organizations.

Increased collaboration with other departments. Contract management is no longer a siloed function. Contract managers will need to collaborate closely with other departments, such as sales, procurement, and legal, to ensure that contracts are aligned with the organization's overall business goals and objectives.

Greater focus on sustainability. Sustainability is becoming an increasingly important issue for organizations, and contract management can play a role in helping organizations achieve their sustainability goals. For example, contract managers can work with suppliers to select sustainable materials and to negotiate terms that promote sustainable practices.

Increased use of artificial intelligence (AI). AI is still in its early stages of development in the context of contract management, but it has the potential to revolutionize the field. For example, AI can be used to automate complex tasks such as contract review and risk assessment.

ABOUT THE AUTHOR

Steven Smith, Ph.d.

Steven Smith is a renowned expert in the field of Construction Management, with a wealth of knowledge and experience spanning both academia and industry. Holding a doctorate in Construction Management, Steven has dedicated his career to advancing the field and contributing to its body of knowledge.

Throughout his academic journey, Steven's passion for understanding the intricacies of construction processes and finding innovative solutions to industry challenges became evident. His doctoral research focused on optimizing project management practices and enhancing productivity in construction projects, leading to a profound understanding of various aspects of construction management and their impact on project success.

BOOKS BY THIS AUTHOR

The Dictionary Of Construction Terminologies: A Compendium Of Knowledge For Students, Academics, Practitioners, And House Owners

The Dictionary of Construction Terminologies

Learn the language of construction from one of the most comprehensive dictionaries of construction terminologies available. From architecture and engineering to materials and equipment, this book covers several aspects of construction terminology in clear and concise language. With thousands of entries, the dictionary is an essential tool for anyone who wants to understand the complex world of construction.

The book features:
Comprehensive coverage of construction terminology
Clear and concise definitions, written in easy-to-understand language
Alphabetical organization for quick and easy reference
It is an essential tool for professionals, students, and anyone interested in the construction field.

Order your copy of the book and start mastering the language of construction!